AS Physics

There's a big jump from GCSE to AS Level Physics.
And with modules to take as early as January, you need to
make sure you hit the ground running.

This book will give you a Head Start — it covers all the AS basics
in enough detail to get you through the first few months, along with
practice questions to make sure you know all the facts.

It's ideal for use in the classroom, or for some extra study.
Make sure you get the grade you deserve.

What CGP is all about

Our sole aim here at CGP is to produce the highest quality books
— carefully written, immaculately presented and dangerously
close to being funny.

Then we work our socks off to get them out to you
— at the cheapest possible prices.

Published by Coordination Group Publications Ltd.

Author:
Richard Tattersall

Design editors:
Chris Dennett
Julie Schofield

ISBN: 978 1 84146 420 6
Groovy website: www.cgpbooks.co.uk
Jolly bits of clipart from CorelDRAW®
Printed by Elanders Hindson Ltd, Newcastle upon Tyne.
Text, design, layout and original illustrations
© Coordination Group Publications Ltd. 2003
All rights reserved.

Contents

Symbols and SI Units i

Section One — Waves

Waves 1
Frequency and Time Period 3
The Wave Equation 4

Section Two — Motion

Distance, Time and Speed 6
Displacement 8
Velocity 10
Acceleration 12
Displacement-Time Graphs 14
Velocity-Time Graphs 16

Section Three — Forces

Forces 18

Section Four — Energy

Kinetic and Gravitational Potential Energy 21
Work 24
Power 26
Efficiency 28

Section Five — Electrostatics

Charge 29
Explaining Electrostatic Phenomena 30

Section Six — Circuit Electricity

Current 32
Voltage 34
Energy in Electrical Circuits 35
Resistance 36
Power 38

Section Seven — Radioactivity

Nuclear Radiation 40
The Random Nature of Radioactive Decay 42

Index 44

Symbols and SI Units

One of the biggest jumps between GCSE Physics and AS Level Physics is in the way things are written down. At AS level, you're expected to start using <u>standard scientific notation</u>...

Standard notation means:
- using the <u>conventional</u> symbols for quantities,
  e.g. a temperature should always have the symbol, T
- writing all quantities in terms of SI units (<u>S</u>tandard <u>I</u>ndex units)
- writing very large and very small numbers in standard form (e.g. 10^6 or 10^{-3})

Those last two points mean you won't have to worry about meeting nasty units like microseconds (μs) and megajoules (MJ) in exams — since they aren't really SI. They'll be written as 10^{-6} s and 10^6 J instead. (If you can't remember how standard index form works, look it up in one of our maths guides... or phone a friend)

The table below lists the different quantities you'll meet in this book, with their standard symbols and units.

Quantity	Symbol	Unit Name	Unit Symbol
Displacement (distance)	s	metre	m
Time	t	second	s
Velocity (speed)	v	metre per second	ms^{-1}
Acceleration	a	metre per second squared	ms^{-2}
Mass	m	kilogram	kg
Force	F	newton	N
Gravitational field strength	g	newton per kilogram	Nkg^{-1}
Energy	E	joule	J
Power	P	watt	W
Frequency	f	hertz	Hz
Wavelength	λ	metre	m
Temperature	T	kelvin	K
Charge	Q	coulomb	C
Electric current	I	amp	A
Potential difference	V	volt	V
Resistance	R	ohm	Ω

At A Level, units like m/s are written ms^{-1}. This is just <u>index notation</u>. (If it doesn't make sense to you, look up 'rules of indices' in a Maths book.)

If you write the unit out in full, it should be all lower case — always.

If the unit comes from someone's name, the symbol should start with a capital letter e.g. Hz.

Here's the unit prefixes for those long numbers:

Multiple	Prefix	Symbol
10^{12}	tera	T
10^9	giga	G
10^6	mega	M
10^3	kilo	k
10^{-2}	centi	c
10^{-3}	milli	m
10^{-6}	micro	μ
10^{-9}	nano	n
10^{-12}	pico	p
10^{-15}	femto	f

Of course, what they don't tell you at A Level is that people in the real world don't use standard scientific notation. A lot of text books that you'll be using will stick to the old prefix system for units.

And if you take Physics at University, that's a whole new ball-game. You'll meet odd units like parsecs and angstroms, which are definitely not SI.

Section One — Waves

Waves

What Are Waves?

1) Waves carry vibrations and energy from one place to another without transferring matter.
2) The motion of turns of a slinky spring can be wavelike.
3) The vibrations and energy you put in by shaking one end are transferred along the spring.

Transverse Waves

Transverse waves have vibrations at 90° to the direction of travel of the wave.

E.g. Shaking a slinky spring from side to side.

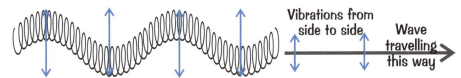

Longitudinal Waves

Longitudinal waves have vibrations along the same direction as the wave is travelling.

E.g. Plucking a slinky spring.

Displacement-Distance Graphs

For any wave we can draw how far each part of the wave is displaced from its equilibrium position for different distances along the wave.

E.g. how far each turn of a slinky spring is displaced sideways as a transverse wave passes along the spring.

(The dashed line shows the displacement of each part of the wave a short time later).

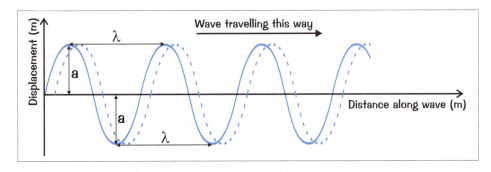

Amplitude (symbol, a) = the furthest displacement of one part of the wave from its equilibrium position, measured in metres.

Wavelength (symbol, λ) = the shortest distance between two parts of the wave that are in the same stage of motion — in particular, the distance between adjacent peaks or troughs, measured in metres.

Waves

Displacement-Time Graphs

We can also consider just one part of the wave (such as one turn of a spring), and plot how its displacement changes with time.

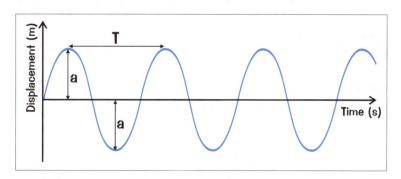

Time period (symbol, T) = the time for one complete oscillation of one part of the wave, measured in seconds.

Have a go at these questions:

1) Sketch a graph of the displacement against distance for two complete wavelengths of a wave of amplitude 0.2 metres and wavelength 1.5 metres.

2) Sketch a graph of the displacement against time for two complete oscillations of one part of a wave of amplitude 0.6 metres and time period 2 seconds.

3) Sketch a graph of the displacement against distance for five full wavelengths of a wave with amplitude 0.01 metres and wavelength 0.02 metres.

4) Sketch a graph of the displacement against time for five complete oscillations of one part of a wave of amplitude 0.05 metres and time period 0.8 seconds.

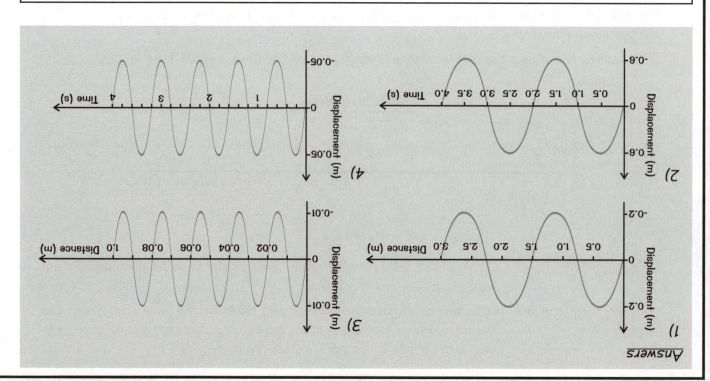

Answers

Section One — Waves

Frequency and Time Period

Frequency and Time Period

Consider one part of a wave (e.g. one turn of a slinky spring) that is vibrating.

If it has a time period of 0.2 seconds, i.e. it takes 0.2 seconds to complete one full oscillation, then in one second it will complete 5 full oscillations.

We say it has a <u>frequency</u> of 5 hertz.

> The number of oscillations of one part of a wave every second is called the <u>frequency</u> of the wave. It has the symbol f and is measured in hertz (symbol, Hz).

1) If we know the time period T then we can work out the frequency using the equation f = 1/T.
2) If we know the frequency f then we can work out the time period using the equation T = 1/f.

Consider the following examples:

1) One coil of a spring oscillates with a time period of 0.008 seconds. What is the frequency of the wave passing along that spring?

   f = 1/T = 1/0.008s = 125Hz

2) A wave has a frequency of 350 Hz. What is the period of oscillation of one part of that wave?

   T = 1/f = 1/350Hz = 0.0029s

Now you try these questions:

1) Ripples on the surface of a pond have a frequency of 12 Hz. What is the time period of oscillation of particles in the water?
2) One turn of a slinky spring takes 0.45 seconds to complete one full oscillation. What is the frequency of the wave on the spring?
3) A radio signal has a frequency of 8×10^5 Hz (800 kHz). What is the time period of oscillations of the electromagnetic field?
4) Oscillations in a sound wave have a time period of 0.002 seconds. What is the frequency of the sound?

Answers
1) 0.083s
2) 2.2Hz
3) 1.25×10^{-6} s
4) 500Hz

Section One — Waves

The Wave Equation

The Wave Equation Relates Speed, Frequency and Wavelength

For a wave of frequency f (in hertz), wavelength λ (in metres) and wave speed v (in metres per second) the wave equation reads:

$$v = f \times \lambda$$

In other words:

$$speed\ (ms^{-1}) = frequency\ (Hz) \times wavelength\ (m)$$

Look at these examples of using the wave equation:

1) Sound is a longitudinal wave. If a sound has a frequency of 250 Hz and a wavelength of 1.32 metres, what is the speed of sound in air?

   v = f x λ, so v = 250Hz x 1.32m = 330ms^{-1}

2) All electromagnetic waves travel at 3×10^8 ms^{-1} in free space. If a radio signal has a wavelength of 1.5 kilometres, what is its frequency? (Hint: radio waves are a member of the electromagnetic spectrum).

   v = f x λ, dividing both sides by λ gives v/λ = f, so f = v/λ = (3×10^8 ms^{-1})/1500m = 2×10^5Hz = 200kHz.

3) If a wave has speed 50 ms^{-1} and frequency 0.8 Hz, what is the wavelength?

   v = f x λ, dividing both sides by f gives v/f = λ, so λ = v/f = (50ms^{-1})/0.8Hz = 62.5m.

Now have a go at these questions:

1) What is the frequency of a water wave of wavelength 0.4 metres and wave speed 0.7 ms^{-1}?
2) What is the wavelength of radio waves of frequency 1×10^8 Hz? (Hint: speed of electromagnetic waves in free space = 3×10^8 ms^{-1})
3) What is the speed of a wave of frequency 800 Hz and wavelength 2.5 metres?
4) What is the frequency of a sound wave of wavelength 0.25 metres? (Hint: speed of sound in air = 330ms^{-1})
5) What is the speed of a wave along a spring if it has frequency 3 Hz and wavelength 1.4 metres?
6) What is the wavelength if the wave speed is 150 ms^{-1} and the frequency is 600 Hz?

Answers
1) 1.75Hz
2) 3m
3) 2000ms^{-1}
4) 1320Hz
5) 4.2ms^{-1}
6) 0.25m or 25cm

Section One — Waves

The Wave Equation

More Questions on Waves

Now have a go at these questions:

Below are graphs of displacement against distance along the wave and displacement of one part of the wave against time for two different waves. In each case:

(a) What is the amplitude of the wave?
(b) What is the wavelength of the wave?
(c) What is the time period?
(d) Calculate the frequency.
(e) Work out the wave speed.

Wave 1

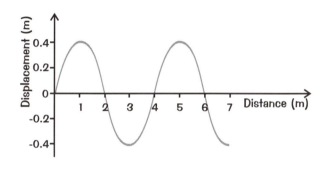 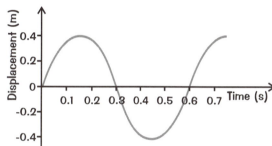

Wave 2

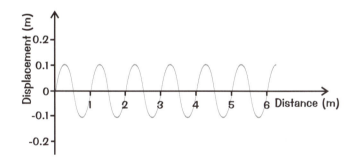 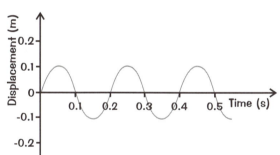

Answers

Wave 1
(a) 0.4m
(b) 4m
(c) 0.6s
(d) 1.67Hz
(e) 6.67ms^{-1}

Wave 2
(a) 0.1m
(b) 1m
(c) 0.2s
(d) 5Hz
(e) 5ms^{-1}

SECTION TWO — MOTION

Distance, Time and Speed

Distance, Time and Speed

Consider the points A and B below. They are separated by a distance which we measure in metres. Now imagine a spider walking from A to B — we can measure the time it takes, in seconds, for it to travel this distance.

A   B

We can then work out the average speed of the spider between these two points using the following simple equation:

> speed (ms^{-1}) = distance travelled (m) / time taken (s)

This is a very useful equation, but it does have a couple of limitations:

1) It only tells you the average speed. The spider could have varied its speed from fast to slow and even gone backwards. So long as it got from A to B in the same time we get the same answer. The average speed is the speed which, if maintained for the whole journey, would take you the given distance in the given time.

2) We assume that the spider takes the shortest possible path between the two points (a straight line), rather than meandering around.

Look at these examples:

> It is important to remember that you should always convert distances to metres, times to seconds and speeds to metres per second when doing calculations involving this equation!

    E.g. A time of 12 minutes is 12 x 60 s = 720 s.
         A distance of 30km is 30 x 1000m = 30 000 m.
         A speed of 10kms^{-1} is 10 x 1000ms^{-1} = 10 000ms^{-1}.
         A speed of 10 metres per minute is 10 / 60ms^{-1} = 0.17ms^{-1}.

1) A car travels 100 metres in 5 seconds. What is its average speed?

   speed = distance/time, so speed = 100m/5s = 20ms^{-1}

2) A train travelling at an average speed of 50ms^{-1} takes 30 minutes to travel between stations. How far apart are the two stations?

   speed = distance/time, multiplying both sides by time gives speed x time = distance.
   30 minutes = 30 x 60s = 1800s.
   So, distance = speed x time, distance = 50ms^{-1} x 1800s = 90 000 m.

3) A spider walks away from you in a straight line. It starts at a point 10cm away, and finishes 50cm away from you. It walks at an average speed of 0.1ms^{-1}. How long does it take?

   Distance travelled = 50cm − 10cm = 40cm = 0.4m.
   From example 2, distance = speed x time, dividing both sides by speed gives:
   distance/speed = time. So time = 0.4m/(0.1ms^{-1}) = 4s.

Distance, Time and Speed

Now have a go at these questions:

1) A cricket ball is thrown 40m at a speed of 18ms^{-1}. How long does it take?
2) A sprinter runs the 100m in a time of 10.5s, what is his average speed?
3) A walker travels at an average speed of 0.5ms^{-1} for half an hour, how far do they walk?
4) A cyclist travels 10km at an average speed of 8ms^{-1}. How long does it take her?
5) A ferry crosses a 250m wide river in 2 minutes. What is its average speed?
6) The speed of light is 3×10^8 ms^{-1}. If it takes light from the sun about 8 minutes to reach us, what is the approximate distance from the earth to the sun?
7) A bird flies 15m between trees in 5s. What is its average speed?
8) How long will it take a horse to gallop 300m across a field at an average speed of 15ms^{-1}?
9) A spaceship flies 3000km in 1 minute. What is its average speed?
10) (difficult!) A blade of grass grows at an average speed of 1 metre per year. How far does it grow in one week?

Answers

1) 2.2 s
2) 9.52 ms^{-1}
3) 900 m
4) 1250 s or 20 mins and 50 s
5) 2.1 ms^{-1}
6) 1.44×10^{11} m
7) 3 ms^{-1}
8) 20 s
9) 5×10^4 ms^{-1} or 50 kms^{-1}
10) 0.019 m = 1.9 cm (speed = $1/3.1536 \times 10^7$ ms^{-1}, time = 6.048×10^5 s, so distance = 0.019 m)

Displacement

Displacement — a Vector Quantity

In order to get from point A to point B, knowing the distance you need to travel is not enough, you must also know the direction you need to travel in. This information, distance plus direction, is known as the <u>displacement</u> from A to B and has the symbol, s. It is a vector quantity since all vectors have both a size and a direction.

Representing Displacement — Scale Drawings

The simplest way to draw a displacement is to draw an arrow — the length of the arrow tells you the distance, and the way the arrow points shows you the direction.

We can do this even for very large displacements so long as we scale down.

For example, a displacement of 3 metres upwards could be represented by an arrow of length 3 centimetres. Using this same scale (1 cm to 1 m) a displacement of 7 metres to the right would be an arrow of length 7 centimetres.

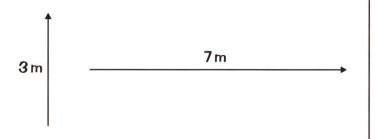

Addition of Two Displacements

We can't simply add together the two distances as this does not account for the different directions of the displacements. What we do is:

1) Draw arrows representing the two vectors.
2) Place the arrows one after the other "tip-to-tail".
3) Draw a third arrow from start to finish. This is your total displacement.

For example, consider adding a displacement of 4 metres to the right to one of 3 metres upwards (using a scale of 1cm to 1m):

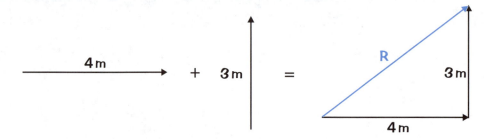

R is called the resultant, it is the sum of the two displacements. You can find the size of R either by measuring the arrow and scaling up or by using Pythagoras. In this case it is 5m in length.

Section Two — Motion

Displacement

Have a go at these questions:

1) Draw arrows representing the following displacements to the given scale.
   - (a) 3 miles upwards (1cm to 1 mile)
   - (b) 12m to the right (1cm to 2m)
   - (c) 4mm downwards (1cm to 1mm)
   - (d) 3.5km northeast (1cm to 1km)
   - (e) 5cm southwest (1cm to 1cm)
   - (f) 24m at a bearing of 030° (1cm to 5m)
   - (g) 110 miles at a bearing of 210° (1cm to 20 miles)
   - (h) 9mm at a bearing of 330° (1cm to 2mm)
   - (i) 18km northwest (1cm to 5km)
   - (j) 9 miles to the left (1cm to 2 miles)

2) Find the lengths of the following displacements by drawing the arrows "tip-to-tail"
   - (a) 5m right and 12m up
   - (b) 8m up and 4m left
   - (c) 6cm right and 8cm left
   - (d) 15 miles down and 20 miles left
   - (e) 3mm left and 12mm right
   - (f) 7.5cm up and 9.5cm right
   - (g) 7km down and 24km right
   - (h) 20m left and 30m down
   - (i) 40 miles left and 50 miles right
   - (j) 60mm up and 20mm right

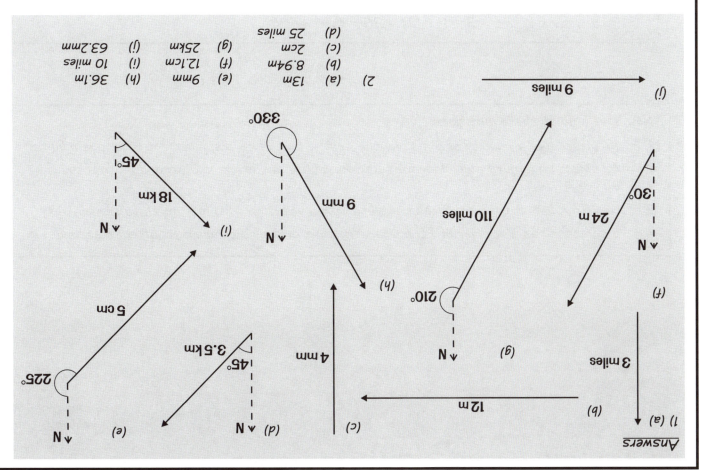

Answers

1) (a) 3 miles ... (see diagrams)

2) (a) 13m (b) 8.94m (c) 2cm (d) 25 miles (e) 9mm (f) 12.1cm (g) 25km (h) 36.1m (i) 10 miles (j) 63.2mm

Section Two — Motion

Velocity

Another quantity that you will have to get used to using is <u>velocity</u>, symbol v.

The Relationship Between Displacement and Velocity

The velocity of an object is given by the following equation:

> velocity (ms^{-1}) = displacement (m) / time taken (t)

Or, in symbols:

> v = s / t

This equation is very similar to the one relating speed and distance (page 6), except that it includes information about the direction of motion.

Consider these examples:

1) An object is displaced 100 metres to the right in a time of 4 seconds. What is its velocity?

    v = s/t, so v = 100m/4s = 25ms^{-1} to the right.
    Note that we must quote the direction as well as the speed.

2) An object has a velocity of 3 metres per second downwards. What is its displacement after 12 seconds?

    v = s/t, multiplying both sides by t gives, v × t = s,
    i.e. s = v × t, so s = 3ms^{-1} × 12s = 36m downwards.

Now have a go at these questions:

1) An object has a velocity of 3 ms^{-1} to the west. What is its displacement after one minute?
2) An object undergoes a displacement of 0.2 metres to the left in 5 seconds. What is its velocity?
3) How long does it take an object travelling with a velocity of 50 ms^{-1} north to travel 1km?
4) If someone has a velocity of 7.5 ms^{-1} south, what is their displacement after 15 seconds?

Answers
1) 180m west.
2) 0.04ms^{-1} left.
3) 20 s.
4) 112.5m south.

Section Two — Motion

Velocity

Velocity is Another Vector Quantity

Just as displacement is distance and direction, so velocity is the <u>speed</u> of an object and the <u>direction</u> it is travelling in.

Again, we can represent velocities with arrows, but now the longer the arrow the greater the speed of the object. A typical scale might be 1cm to 1ms^{-1}.

For example, the two velocities of 5 metres per second to the right and 3 metres per second downwards might be drawn (with a scale of 1cm to 1ms^{-1}):

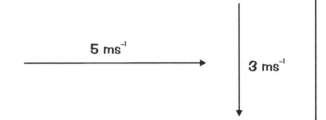

Changes in Velocity

change in velocity (ms^{-1}) = final velocity (ms^{-1}) − initial velocity (ms^{-1})

This is not quite as simple as it looks. We cannot just take one speed from the other, we have to account for the directions of the two velocities.

Consider the example of: an initial velocity = 5ms^{-1} to the right, and a final velocity = 3ms^{-1} down.

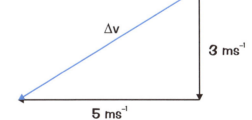

Δv is the <u>change in velocity</u>.

We found Δv by reversing the arrow for the initial velocity and adding this new arrow to the final velocity.

Δv = 5.8ms^{-1} diagonally down and left.

Have a go at finding the changes in velocity in these cases:
1) initial velocity = 10ms^{-1} right; final velocity = 5ms^{-1} left.
2) initial velocity = 4ms^{-1} up; final velocity = 4ms^{-1} right.
3) initial velocity = 3ms^{-1} down; final velocity = 4ms^{-1} left.

Answers
1) 15ms^{-1} to the left.
2) 5.7ms^{-1} diagonally down and right.
3) 5ms^{-1} diagonally up and left.

Acceleration

Acceleration — the Change in Velocity Every Second

Acceleration (in metres per second^2) = $\dfrac{\text{change in velocity of an object (in metres per second)}}{\text{time taken (seconds)}}$

So, Acceleration (ms^{-2}) = $\dfrac{\text{final velocity (ms}^{-1}) - \text{initial velocity (ms}^{-1})}{\text{time taken (s)}}$

Or in symbols: $a = \dfrac{v - u}{t} = \dfrac{\Delta v}{t}$

For simplicity we will only worry about velocities in <u>one dimension</u>, say left to right.

This has the advantage that we don't need to bother drawing out all the arrows for the velocities.

But we still need to worry about the difference between velocities from right to left and velocities from left to right.

We must choose a direction to be positive — let's say right. All velocities in this direction will from now on be positive, and all those in the opposite direction (left) will be negative.

Look at these examples to see how this works:

1) A car starts off moving to the right at 10 metres per second. After 20 seconds it is moving to the left at 5 metres per second. What was its acceleration during this time?
   u = 10ms^{-1} to the right = +10ms^{-1}
   v = 5ms^{-1} to the left = -5ms^{-1}
   So, a = (v − u)/t = (-5ms^{-1} − 10ms^{-1})/20s = (-15ms^{-1})/20s = -0.75ms^{-2}.
   The acceleration is negative so it is to the left.

2) An object accelerates from rest at 5ms^{-2} to the right. If its final velocity is 20ms^{-1} to the right, how long has it been accelerating for?
   u = 0
   v = 20ms^{-1} to the right = +20ms^{-1}
   a = (v − u)/t, multiplying both sides by t gives t × a = v − u, dividing both sides by a gives t = (v − u)/a.
   So, t = (20ms^{-1} − 0)/(5ms^{-2}) = 4s.

Have a go at these questions:

1) A train has an initial velocity of 12 ms^{-1} to the left. After 20 seconds it is moving to the right at 18 ms^{-1}. What was its acceleration during this time?
2) A ship accelerates at a uniform rate of 0.1 ms^{-2} to the right. If its initial velocity is 1.5 ms^{-1} to the right and its final velocity is 4 ms^{-1} in the same direction, how long has it been accelerating for?
3) True or false? Acceleration is a vector.

Answers
1) 1.5ms^{-2} to the right
2) 25s
3) True

Section Two — Motion

Acceleration

Falling — the Acceleration Due to Gravity

When an object is dropped it accelerates downwards at a constant rate of roughly 10 ms^{-2}. This is the <u>acceleration due to gravity</u>.

It seems sensible to take the upward direction as positive and down as negative, making the acceleration due to gravity <u>-10ms^{-2}</u>.

Look at the following examples:

1) What is the vertical velocity of a skydiver 5 seconds after jumping out of a plane? (Ignore the skydiver's horizontal motion)

   $u = 0$
   $a = -10ms^{-2}$
   Example 2 on page 12 gives $t \times a = v - u$,
   so adding u to each side gives $v = u + (t \times a)$.
   So $v = 0 + (5s \times -10ms^{-2}) = 0 - 50ms^{-1}$
   $= -50ms^{-1} = 50ms^{-1}$ down.

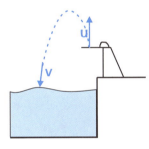

2) A diver jumps upwards off a springboard. After 2 seconds he hits the water travelling downwards at 18ms^{-1}. What was his initial velocity?

   $v = 18ms^{-1}$ down $= -18ms^{-1}$
   $a = -10ms^{-2}$
   From example 1, $v = u + (t \times a)$, subtracting $(t \times a)$ from each side gives $v - (t \times a) = u$.
   So, $u = -18ms^{-1} - (2s \times -10ms^{-2}) = -18ms^{-1} - (-20ms^{-1})$
   $= -18ms^{-1} + 20ms^{-1} = 2ms^{-1} = 2ms^{-1}$ upwards.

Try these. (Hint: it is often useful to draw a little diagram of what is going on in these questions)

1) An apple falls from a tree and hits the ground at 2 ms^{-1}. For how long has it been falling?
2) A pea dropped from the top of a tall building falls for 3 seconds. Ignoring air resistance, with what velocity does it hit the ground?
3) A stone is thrown downwards. It hits the ground at 15 ms^{-1} after 0.7 seconds. With what velocity was it thrown?
4) A package is dropped from a stationary helicopter. Ignoring air resistance, with what velocity does it hit the ground 10 seconds later?
5) A sandbag is dropped from a stationary hot air balloon and it hits the ground at a velocity of 50 metres per second downwards. How long has it been falling for?
6) A ball is thrown upwards. After 2 seconds it is caught moving downwards at 10 ms^{-1}. With what velocity was it thrown upwards?

Answers
1) 0.2s
2) 30ms^{-1} down
3) 8ms^{-1} down
4) 100ms^{-1} down
5) 5s
6) 10ms^{-1} up

Displacement-Time Graphs

Drawing Graphs to Show How Far Something Has Travelled

A graph of displacement against time tells you <u>how far</u> an object is from a given point, in a given direction, as time goes on. As it moves away from that point the displacement on the graph goes up and as it moves towards it the displacement goes down:

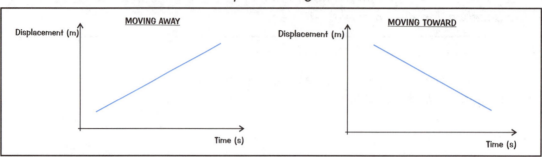

Importantly, these graphs only tell you about motion in <u>one dimension</u>, so for example, they can tell you how far up a ball has been thrown, but not how far it has moved horizontally.
We can also use these graphs to calculate the velocity of an object (in the given direction).

Consider the example below, it shows the displacement-time graph for a car accelerating to a constant speed and then braking suddenly.

We can read the following <u>directly</u> off the graph:

1) It took 20 seconds to accelerate to full speed.
2) It travelled 100 metres in that time.
3) It travelled at constant velocity for the next 10 seconds.
4) It travelled 200 metres in that time.
5) It took 5 seconds to stop fully.
6) It travelled 50 metres in that time.
7) It remained stationary at a displacement of 350 metres from its starting point.

We can work out three more details of the car's journey:

1) The value of the <u>constant velocity</u> it had between 20 and 30 seconds.
2) Its <u>average velocity</u> for the whole journey.
3) Its <u>average speed</u> for the whole journey.

When an object is travelling at a <u>steady velocity</u> its displacement-time graph is a <u>straight line</u>, with a gradient equal to the velocity.

velocity (ms^{-1}) = gradient = $\dfrac{\text{change in distance travelled (m)}}{\text{change in time (s)}}$ = $\dfrac{300m - 100m}{30s - 20s}$ = $\dfrac{200m}{10s}$ = $\boxed{20ms^{-1}}$

To calculate the average speed for the whole journey we use the formula:

average velocity (ms^{-1}) = $\dfrac{\text{total displacement (m)}}{\text{total time taken (s)}}$ = $\dfrac{350m}{35s}$ = $\boxed{10ms^{-1}}$

and, average speed (ms^{-1}) = $\dfrac{\text{total distance travelled (m)}}{\text{total time taken (s)}}$ = $\dfrac{350m}{35s}$ = $\boxed{10ms^{-1}}$

In this case, the average speed is the same as the average velocity, because the car doesn't change <u>direction</u>. The total distance is the +ve displacement plus the -ve displacement.

Displacement-Time Graphs

Have a go at analysing these graphs:

Write down as much as you can about the motion of the objects represented by the following graphs. Work out any steady velocities, the average velocity and average speed for the journey.

Graph 1:

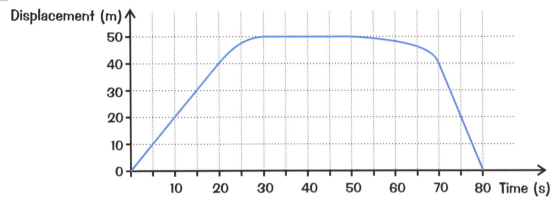

Graph 2:

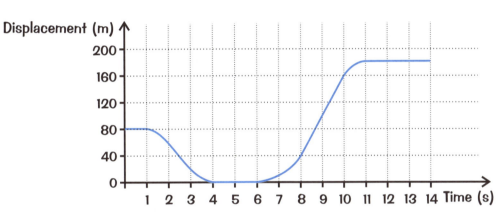

Answers

Graph 1:

Readings from the graph:
1) Object travels at a steady speed for 20 seconds.
2) It covers 40 metres in that time.
3) It then decelerates to zero speed in 10 seconds.
4) It covers 10 metres in this time.
5) It remains stationary at a distance of 50 metres for the next 20 seconds.
6) Object then accelerates back towards its starting point for 20 seconds.
7) It covers 10 metres in this time.
8) It continues at a steady speed for 10 seconds.
9) It travels 40 metres in this time. Returning to its starting point.

Calculations:
1) Original steady velocity = (+40m − 0)/(20s − 0)
   = +40m/20s = +2ms^{-1}
2) Returning steady velocity = (0 − 40m)/(80s − 70s)
   = −40m/10s = −4ms^{-1}
3) Average velocity = (final displacement − initial displacement)/time = (0m − 0m)/80s = 0ms^{-1}
4) Average speed = (distance travelled in +ve direction + distance travelled in −ve direction)/time
   = (50m + 50m)/80s = 1.25ms^{-1}

Graph 2:

Readings from the graph:
1) Remains stationary at a distance of 80m for 1 second.
2) Accelerates back towards zero metres for 1 second.
3) It covers 20 metres in this time.
4) Object travels at a steady speed towards zero metres for 1 second.
5) It covers 40 metres in this time.
6) It decelerates for 1 second until it is stationary.
7) It travels 20 metres in this time.
8) It remains at zero metres for 2 seconds.
9) Object accelerates away for 2 seconds.
10) It travels 40 metres in this time.
11) It continues at a steady speed for 2 seconds.
12) It covers 120 metres in this time.
13) It decelerates for 1 second until it stops.
14) It covers 20 metres in this time.
15) Object remains stationary at a distance of 180 metres for the next 3 seconds.

Calculations:
1) First constant velocity = (20m − 60m)/(3s − 2s)
   = −40m/1s = −40ms^{-1}
2) Second constant velocity = (160m − 40m)/(10s − 8s)
   = 120m/2s = 60ms^{-1}
3) Average velocity = (180m − 80m)/14s = 100m/14s
   = 7.14ms^{-1}
4) Average speed = (180m + 80m)/14s = 260m/14s
   = 18.6ms^{-1}

Velocity-Time Graphs

Drawing Graphs to Show the Speed of an Object

Clearly, we can also draw graphs that show the velocity of an object moving in one dimension.

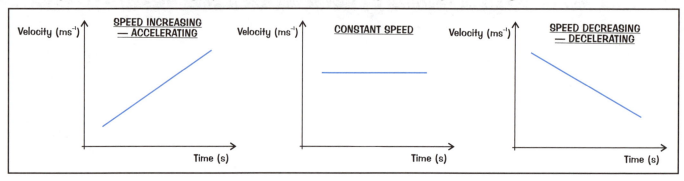

We can use a velocity-time graph to calculate two things:
1) The distance the object has moved.
2) The acceleration.

Calculating the Distance Travelled

To find the distance an object travels between two times:
1) Draw vertical lines up from the horizontal axis at the two times as shown.
2) Work out the "area" of the shape formed by these lines.
3) Remember that although we call it an area we are actually multiplying time (the horizontal length) by average speed (the average vertical length) so the result is a distance.

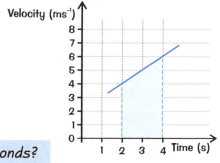

E.g. What is the distance travelled between 2 seconds and 4 seconds?

The shape is a trapezium, so the area = ½(a + b)×h = ½(4 + 6)×2 = 5ms⁻¹ × 2s = 10m.

Calculating the Acceleration

The acceleration of an object travelling in one dimension is given (see page 12) by:

Acceleration (ms⁻²) = change in velocity (ms⁻¹) / time taken (s)

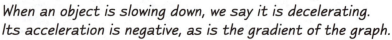

This is just the gradient of the velocity-time graph.

E.g. What is the acceleration between 10 and 20 seconds?

Acceleration = (4ms⁻¹ − 3ms⁻¹)/(20s − 10s)
= 1ms⁻¹ / 10s = 0.1ms⁻²

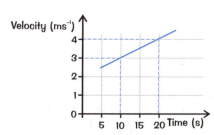

When an object is slowing down, we say it is decelerating. Its acceleration is negative, as is the gradient of the graph.

E.g. What is the acceleration between 5 and 15 seconds?

Acceleration (in ms⁻²) = (10ms⁻¹ − 15ms⁻¹)/(15s − 5s)
= −5ms⁻¹ / 10s = −0.5ms⁻²
or a deceleration of 0.5ms⁻²

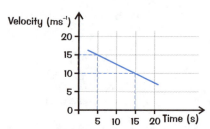

Section Two — Motion

Velocity-Time Graphs

Try these:

1) Calculate the accelerations in each of the three sections of each graph.
2) Calculate the distances travelled in each of the three sections of each graph and calculate the total distance travelled in each case.

a)
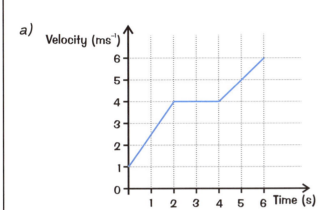

b)
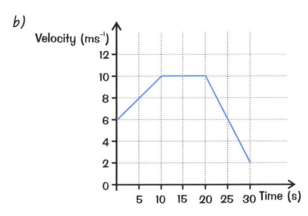

c)
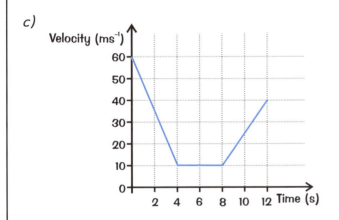

d)
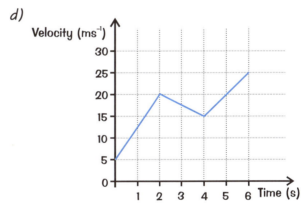

Answers

1)
(a) $1.5ms^{-2}$, $0ms^{-2}$, $1ms^{-2}$
(b) $0.4ms^{-2}$, $0ms^{-2}$, $-0.8ms^{-2}$
(c) $-12.5ms^{-2}$, $0ms^{-2}$, $7.5ms^{-2}$
(d) $7.5ms^{-2}$, $-2.5ms^{-2}$, $5ms^{-2}$

2)
(a) 5m, 8m, 10m, total = 23m
(b) 80m, 100m, 60m, total = 240m
(c) 140m, 40m, 100m, total = 280m
(d) 25m, 35m, 40m, total = 100m

Forces

Newton's Second Law

It is difficult to explain precisely what a "force" is, so instead we talk about what forces do.

Forces stretch things, forces squash things, forces twist things, but most importantly, forces make things go faster (or slower or change what direction they are moving in).

> When a force acts on an object it changes the velocity of the object.

In other words, applying a force to an object makes it <u>accelerate</u>.

This acceleration is <u>directly proportional</u> to the force. This just means that, for the same object, if you double the force applied, you double its acceleration.

We can write down this equation:

> Force (N) = mass of object (kg) × acceleration of object (ms^{-2})

Or, in symbols:

$$F = m \times a$$

This is <u>Newton's second law of motion</u>.

Have a look at this example:

a) A car of mass 1000kg accelerates from rest (0ms^{-1}) to 15ms^{-1} in 20s. What is the force accelerating it?

b) The same car then stops suddenly in 5s. What is the braking force?

a) $v = 15$ms^{-1}
   $u = 0$ms^{-1}
   $t = 20$s
   $a = (v - u)/t$, so $a = (15$ms$^{-1} - 0)/20$s $= 0.75$ms^{-2}.
   Then $F = m \times a = 1000$kg $\times 0.75$ms$^{-2} = 750$N.

b) $v = 0$ms^{-1}
   $u = 15$ms^{-1}
   $t = 5$s
   Again $a = (v - u)/t$, so $a = (0 - 15$ms$^{-1})/5$s $= -3$ms^{-2}
   (the acceleration is negative because the car is slowing down)
   Then $F = m \times a = 1000$kg $\times (-3$ms$^{-2}) = -3000$N
   (the force is negative because it is slowing the car down)

Forces

Finding the Force When Given the Mass and the Acceleration

Try these (they are like the example on page 18):

1) A bus of mass 10 000kg accelerates at 0.25ms^{-2}. What is the force acting on it?
2) A car pulls a caravan of mass 800kg. If it accelerates at 0.4ms^{-2}, what force must the caravan experience?
3) An apple of mass 0.1kg falls with acceleration of 10ms^{-2}.
   What is the gravitational force pulling it down (its weight)?

Finding the Acceleration When Given the Force and the Mass

Consider this example:
   What would be the acceleration of a 500g mass if a force of 10N acted on it?
   $F = m \times a$, dividing both sides by m gives $F/m = a$, so $a = F/m = 10N/0.5kg = 20ms^{-2}$.

Now try these questions:

4) What would be the acceleration of an arrow of mass 0.3kg if the force from the strings in the bow is 200N?
5) What would be the acceleration of a train of mass 10 000 kg if the force from the engine is 8000N?
6) What would be the acceleration of a bullet in a rifle if the bullet has a mass of 0.008kg and the force accelerating it is 2000N?

Finding the Mass When Given the Acceleration and the Force

Consider this example:
   What is the mass of an object if a force of 250N produces an acceleration of 2ms^{-2}?
   $F = m \times a$, dividing both sides by a gives $F/a = m$, so $m = F/a = 250N/(2ms^{-2}) = 125kg$

Now try these questions:

7) What is the mass of a sailing boat if a force of 120N produces an acceleration of 0.5 ms^{-2}?
8) What is the mass of a ship if a force of 50 000N produces an acceleration of 0.2ms^{-2}?
9) What is the mass of a box if a force of 50N produces an acceleration of 8ms^{-2}?

Answers
1) 2500N   2) 320N   3) 1N   4) 667ms^{-2} 5) 0.8ms^{-2} 6) 2.5 × 10^5 ms^{-2} 7) 240kg   8) 2.5 × 10^5 kg   9) 6.25kg

Section Three — Forces

Forces

Balanced and Unbalanced Forces

Force is a <u>vector</u>, just like displacement or velocity.

When more than one force acts on a body, we have to add them together in just the same way as we add displacements or velocities.

We find the resultant force by putting the arrows "tip-to-tail".

If the resultant force is zero we say that the forces are <u>balanced</u>.

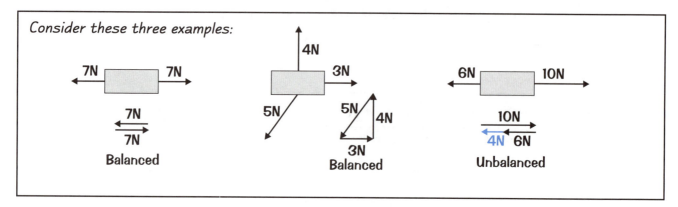

If there is a resultant force then the forces are <u>unbalanced</u>.

We say there is a <u>net force</u> on the object.

Example:

A force of 10 newtons to the right and a force of 6 newtons to the left result in a net force of 4 newtons to the right. If these forces acted on an object of mass 5 kilograms it would produce an acceleration given by:

$$a = F/m = 4N/5kg = 0.8 ms^{-2}$$

Have a go at these:

Work out the net forces on these objects and calculate the acceleration they would cause:

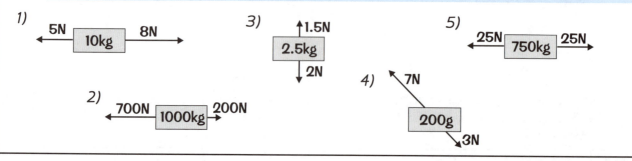

Answers

1) 3N to the right, 0.3ms^{-2} to the right.
2) 500N to the left, 0.5ms^{-2} to the left.
3) 0.5N down, 0.2ms^{-2} down.
4) 4N diagonally up and left, 20ms^{-2} in the same direction.
5) No net force, no acceleration.

Section Three — Forces

SECTION FOUR — ENERGY

Kinetic and Gravitational Potential Energy

Kinetic Energy

Energy is a curious thing. You can't pick it up and look at it, neither can you measure it out in a beaker or flask.

One thing, however, is for certain — if you are moving then you have energy.

This movement energy is more properly known as kinetic energy, and there is a simple formula for working it out:

> If a body of mass m (in kilograms) is moving with speed v (in metres per second) then its kinetic energy (in joules) is given by:
>
> Kinetic Energy = ½ × Mass × Speed²

Or, in symbols:

$$E_k = \tfrac{1}{2} \times m \times v^2$$

This equation involves squaring the speed, which makes the maths a little bit more tricky. Have a look at the following examples, and then try the questions after them.

Examples:

1) A car of mass 800kg is travelling with a speed of 20ms⁻¹. What is its kinetic energy?
   $E_k = \tfrac{1}{2} \times m \times v^2$, so $E_k = \tfrac{1}{2} \times 800\text{kg} \times (20\text{ms}^{-1})^2 = \tfrac{1}{2} \times 800\text{kg} \times 400\text{m}^2\text{s}^{-2}$
   $= 1.6 \times 10^5 \text{J}$.

2) A ball has a speed of 2.5ms⁻¹ and has kinetic energy equal to 0.8J. What is the mass of the ball?
   $E_k = \tfrac{1}{2} \times m \times v^2$, multiplying both sides by 2 gives $2 \times E_k = m \times v^2$, dividing both sides by v^2 gives $(2 \times E_k)/v^2 = m$, so $m = (2 \times E_k)/v^2 = (2 \times 0.8\text{J})/(2.5\text{ms}^{-1})^2$
   $= 1.6\text{J}/(6.25\text{m}^2\text{s}^{-2}) = 0.256\text{kg} = 256$ grams.

3) A bullet has kinetic energy equal to 1200J. If its mass is 0.01kg, what is its speed?
   From example 2) $2 \times E_k = m \times v^2$, dividing both sides by m gives $(2 \times E_k)/m = v^2$, taking square roots of both sides gives $\sqrt{(2 \times E_k)/m} = \sqrt{v^2}$, but $\sqrt{v^2} = v$,
   so $v = \sqrt{(2 \times E_k)/m} = \sqrt{(2 \times 1200\text{J})/0.01\text{kg}} = 490\text{ms}^{-1}$.

Now try these questions:

1) An arrow of mass 0.4kg is travelling at a speed of 80ms⁻¹. What is its kinetic energy?
2) A ship has kinetic energy equal to 6 × 10⁷J when moving at 15ms⁻¹. What is its mass?
3) A snail of mass 0.05kg has kinetic energy 1 × 10⁻⁶J. What is its speed?

Answers:
1) 1280J
2) 5.3 × 10⁵kg
3) 0.006ms⁻¹ or 0.6cms⁻¹

Kinetic and Gravitational Potential Energy

Gravitational Potential Energy

When an object falls its speed increases. As its speed increases so does its kinetic energy.
Where does it get this energy from?
Answer: from the gravitational potential energy it had before it fell.

There is a simple equation for working out the gravitational potential energy an object has:

> If a body of mass m (in kilograms) is raised through a height h (in metres) the gravitational potential energy (in joules) it gains is given by:
>
> Gravitational Potential Energy (J) = mass (kg) × gravitational field strength (Nkg⁻¹) × height (m)

So, in symbols it reads: $E_p = m \times g \times h$

where the gravitational field strength is the ratio between an object's weight and its mass (in newtons per kilogram). It is given the symbol g, and at the surface of the earth has an approximate value of 10Nkg⁻¹.

Consider these examples:

1) An 80 kilogram person in a lift is raised 45 metres. Assuming g = 10Nkg⁻¹ what is the increase in gravitational potential energy?
E_p = mgh, so E_p = 80kg × 10Nkg⁻¹ × 45m = 36 000J

2) A mass raised 10 metres gains gravitational potential energy equal to 50 joules. What is that mass?
E_p = mgh, dividing both sides by gh gives E_p/gh = m,
so m = E_p/gh = 50J/(10Nkg⁻¹ × 10m) = 0.5kg or 500g.

3) 600 kilograms of bricks are given 12 000 joules of gravitational potential energy. Through what height have they been raised?
E_p = mgh, dividing both sides by mg gives E_p/mg = h,
so h = E_p/mg = 12 000J/(600kg × 10Nkg⁻¹) = 2m.

Now try these questions:

1) How much more gravitational potential energy does a 750 kilogram car have at the top of a 300 metre hill compared to an identical car at the bottom of the hill? (Assume g = 10Nkg⁻¹)
2) What mass, when raised through 7 metres gains gravitational potential energy equal to 600 joules? (assume g = 10Nkg⁻¹)
3) A 60 kilogram person gains 25 000 joules of gravitational potential energy, how high have they climbed?

Answers
1) 2.25 × 10⁶J or 2.25MJ
2) 8.6kg
3) 42m

Section Four — Energy

Kinetic and Gravitational Potential Energy

The Conservation of Energy applied to Falling Bodies

The principle of <u>conservation of energy</u> states that:

"Energy cannot be created or destroyed, only converted into other forms"

When an object is <u>falling</u> its gravitational potential energy is converted into kinetic energy, so:

Kinetic Energy Gained (in joules) = Gravitational Potential Energy Lost (in joules)

When an object is <u>thrown</u> or <u>catapulted</u> upwards, its kinetic energy is converted into gravitational potential energy, so:

Gravitational Potential Energy Gained (in joules) = Kinetic Energy Lost (in joules)

This can be very useful in solving problems.
Read through the examples and then have a go at the questions afterwards.

Examples:

1)   An apple of mass 0.1 kilograms falls from a tree of height 2 metres.
     With what speed does it hit the ground?

   E_p lost = mgh = 0.1kg × 10Nkg^{-1} × 2m = 2J
   Therefore E_k gained = 2J, so E_k = ½ × m × v^2 = 2J.
   From page 21, v = √((2 × E_k)/ m), so v = √((2 × 2J)/ 0.1kg) = √(4J/ 0.1kg) = 6.3ms^{-1}

2)   A ball of mass 0.2 kilograms is thrown upwards at 10 metres per second.
     How high does it get before it slows to 0 metres per second?

   E_k lost = ½ × m × v^2 = ½ × 0.2kg × (10ms^{-1})^2 = 10J.
   Therefore, E_p gained = 10J, so E_p = mgh = 10J.
   From page 22, h = E_p/mg, so h = 10J/(0.2kg × 10Nkg^{-1}) = 5m.

Now try these questions:

1)   A book of mass 0.5 kilograms falls off a table 1 metre from the floor. With what speed is it travelling when it lands?
2)   A bullet of mass 0.01 kilograms is fired upwards at 400ms^{-1}. What height does it reach before slowing to 0ms^{-1}?

Answers
1) 4.5ms^{-1} 2) 8000m or 8km

Work

Work — the Amount of Energy a Force gives an Object

When you push an object you can increase its energy by:
1) Pushing it up hill,
2) Accelerating it,
3) Doing both at once.

In any case, the amount of energy (in joules) that a force gives an object is called the work done, and can be calculated using the following formula:

> Work Done by a Force = Size of Force × Distance Moved in the Direction of the Force
> (in joules)   (in newtons)   While the Force is Acting (in metres)

Or, in symbols:

$$W = F \times d$$

Consider the following examples:

1) A 5 newton force to the north pushes an object 3 metres in the same direction. What is the work done?
   $W = F \times d$, so $W = 5N \times 3m = 15J$.

2) A 10 newton force to the north pushes an object 15 metres in a north-easterly direction. What is the work done?
   N.B. We must consider the distance travelled in the direction of the force (i.e. north)

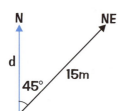

   $d = 15 \cos 45° = 10.6m$
   So, $W = F \times d = 10N \times 10.6m = 106J$.

3) A force of 35 newtons north acts on an object as it moves 7m in a westerly direction. What is the work done?
   The object is moving at 90° to the force, so $d = 0m$.
   Therefore, $W = F \times d = 35N \times 0m = 0J$.

Have a go at these questions:

1) A force of 25 newtons to the west moves an object 40 metres in the same direction. What is the work done?
2) A force of 10 newtons to the north east acts on an object as it moves 25 metres to the south east. What is the work done?
3) A force of 3 newtons to the west acts on an object as it moves 10 metres to the south west. What is the work done?

Answers
1) 1000J
2) 0J
3) 21.2J

Section Four — Energy

Work

Work Done = Increase in Gravitational and Kinetic Energy

There are three possible examples:

1) The work done goes <u>entirely into the gravitational potential energy</u> of an object.
   E.g. if you are lifting an object straight upwards.

   Work done = force × distance
   = weight of object × height lifted
   = mass of object × gravitational field strength × height

   So: <u>work done = mgh = the increase in gravitational energy</u>

2) The work done goes <u>entirely into the kinetic energy</u> of an object.
   E.g. if a 5 newton force acts on a 3 kilogram body over a distance of 10 metres, what is its final speed if it was initially at rest?

   Work done = increase in kinetic energy
   $F \times d = ½ \times m \times v^2$, dividing both sides by m gives:
   $(F \times d)/m = (½ \times m \times v^2)/m$
   So, $(F \times d)/m = ½ \times v^2$, multiplying both sides by 2 gives:
   $2 \times (F \times d)/m = v^2$, finally, taking the square root of both sides gives:
   $v = \sqrt{(2 \times (F \times d)/m)} = \sqrt{(2 \times (5N \times 10m)/3kg)} = \sqrt{(100/3)} = \underline{5.8ms^{-1}}$

3) The work done goes into increasing <u>both the kinetic and the gravitational energy</u>.

   Work done = increase in E_k + increase in E_p
   $F \times d = ½ \times m \times v^2 + mgh$

Have a go at these questions:

1) A 100 newton force lifts a 5 kilogram object 2 metres. When the force is removed, the object continues to move upwards. Calculate (a) the work done by the force; (b) the gain in gravitational potential energy (using $g = 10Nkg^{-1}$); (c) the gain in kinetic energy.
2) A 10 newton force pushes an object of mass 2 kilograms horizontally on a frictionless surface for 25 metres. Calculate (a) the work done; (b) the final speed of the object if it was initially at rest.

Answers
1) (a) 200J
   (b) 100J
   (c) 100J
2) (a) 250J
   (b) 15.8ms^{-1}

Section Four — Energy

Power

Power — the Work Done every Second

In mechanical situations, whenever energy is converted, work is being done.

For example, when an object is falling, the force of gravity is doing work on that object equal to the increase in kinetic energy.

The rate at which this work is being done is called the <u>power</u>.

We can write the equation:

> Power (in watts) = Work Done (in joules) / Time Taken (in seconds)

Or, in symbols:

$$P = W / t$$

Consider the following examples:

1) If 10 joules of work is done in 2 seconds, what is the power?

    $P = W/t = 10J/2s = 5W.$

2) A force of 100 newtons pushes an object 5 metres in 4 seconds. What is the power? (Assume the motion is in the same direction as the force.)

    $W = F \times d = 100N \times 5m = 500J$

    $P = W/t = 500J/4s = 125W.$

3) For how long must a 5 kilowatt (5000 watt) engine run to do 200 kilojoules (2×10^5 joules) of work?

    $P = W/t$, multiplying both sides by t gives $P \times t = W$, dividing both sides by P gives: $t = W/P.$

    So, $t = W/P = 2 \times 10^5 J / 5000W = 40s.$

Try these:

1) If an elevator mechanism works at 15 kilowatts, how long does it take to do 100 kilojoules of work?
2) What is the power output of a motor if it does 250 joules of work in 4 seconds?
3) If a force of 200 newtons pushes an object 1.5 kilometres in a minute, at what power is it working? (Assume the motion is in the same direction as the force.)

Answers
1) 6.7s
2) 62.5W
3) 5000W

Section Four — Energy

Power

The Power Developed by a Moving Force

We can derive a useful equation for the work being done by a force every second on a moving object. Follow through the working below:

For example, at what power is a car engine working if it is producing a driving force of 2000 newtons and moving at a steady speed of 30 metres per second?

In symbols, the power is given by: $P = W/t$

But, $W = F \times d$, so we can substitute for the work done giving: $P = (F \times d)/t$

Now, $(F \times d)/t$ is the same as $F \times (d/t)$ so: $P = F \times (d/t)$

Finally we use the fact that d/t = distance travelled/time taken = the speed, v.

So, $P = F \times (d/t) = F \times v$

Power (in watts) = Force (in newtons) × Speed (in metres per second)

For our example, $P = 2000N \times 30ms^{-1} = 60\,000W = \underline{60kW}$

Note that all the above holds true only when the object is moving at a <u>constant velocity</u> in the <u>same direction as the force</u>.

Have a go at these:

1) What is the power developed by a train engine if the driving force is 1.8×10^5 newtons and the speed is 40 metres per second?
2) A skydiver is falling at a terminal velocity of 45 metres per second. If her weight is 700 newtons, at what rate is gravity doing work on her?

Answers
1) 7.2×10^6W, or 7.2MW
2) 31 500W, or 31.5kW

Section Four — Energy

Efficiency

How Much of what you Put In do you Get Out?

For most mechanical systems you put in energy of one form and it gives out energy in another.

However, some energy is converted into forms that are not useful.

For example, you put energy into a motor in the form of electrical energy and as well as the kinetic energy of the motor spinning some energy is converted into heat and sound.

We measure the efficiency of a system by the _percentage of total energy put in that is converted to useful forms_.

As an example consider raising a load using a pulley

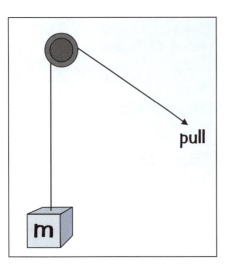

The _energy you put in_ is the work you do pulling the rope.

The _useful energy out_ is the gravitational potential energy gained by the load.

Some energy is converted into heat and sound by _friction_ at the pulley.

If the mass of the load is 10 kilograms and it is raised 5 metres, and you pull with a force of 120 newtons for the same 5 metres, we can work out the efficiency of the pulley.

(Take $g = 9.81 Nkg^{-1}$)

Energy in = Work done = F × d = 120N × 5m = __600J__

Useful energy out = Potential energy gained = mgh = 10kg × 9.81Nkg⁻¹ × 5m = __490.5J__

Efficiency = (Useful energy out/Total energy in) × 100

= (490.5J/600J) × 100 = __82%__

Have a go at these questions:

1) A motor uses 300 joules of electrical energy in lifting a 10 kilogram mass through 2 metres. What is its efficiency? (Take $g = 9.81 Nkg^{-1}$)

2) It takes 800 kilojoules (8×10^5 joules) of chemical energy from the petrol in a car engine to accelerate a 500 kilogram vehicle to 20 metres per second on a flat road:
(a) What is the gain in kinetic energy of the car?
(b) What is the efficiency of the engine?

Answers
1) 65.4%
2) (a) 1×10^5 J, or 100kJ.
    (b) 12.5%

Section Four — Energy

SECTION FIVE — ELECTROSTATICS

Charge

Some Particles have a Property called Charge

All the particles you will have met at GCSE have mass. Because of this mass they all attract each other. This force of attraction is called gravity. Some particles also have a property that we call charge, measured in coulombs, symbol C. Importantly, this charge can be either positive or negative, and the force between charged particles can be an attraction or a repulsion.
This force is called the electrostatic force and has the following properties:

1) Particles with <u>opposite</u> charges <u>attract</u>.
2) Particles with the <u>same</u> charge <u>repel</u>.
3) The force is bigger if the sizes of the charges (in coulombs) are bigger, or if the particles are closer together.

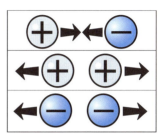

Have a look at these worked examples.

QUESTIONS:
1) Which of these pairs of particles attract and which repel?
2) Between which pair is the electrostatic force greatest?

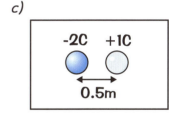

ANSWERS:
1) a) attract; b) repel; c) attract
2) c) because they all have the same sized charges, but these are closer together

Have a go at these questions:

1) Which of these pairs of particles attract and which repel?
2) Between which pair is the electrostatic force greatest?

a)
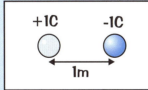

b)
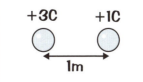

c)
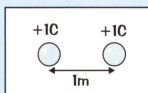

Answers:
1) a) attract; b) repel; c) repel
2) b) because they are all the same distance apart, but one of these charges is greater

Explaining Electrostatic Phenomena

What Happens When Charged Particles Collide?

1) When a charged particle is fired towards another, the electrostatic force between them has an effect on the trajectory of the particle.
2) The diagrams below show the paths of a negatively charged particle fired at high speed towards a fixed positive charge:

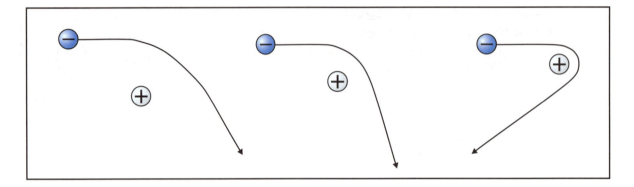

Importantly:

> The force gets _bigger_ the _closer_ together the two particles are.

This means that the closer the negative particle is to the positive one, the more it gets deflected.

Have a go at making a prediction:

1) Predict the paths for a positive charge fired at high speed towards a fixed positive charge:

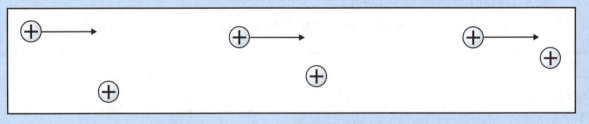

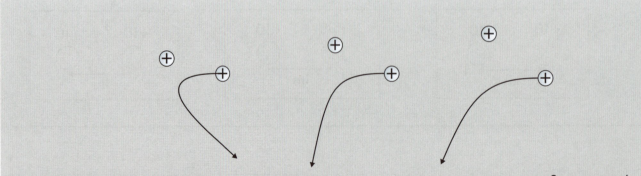

Answers

1) The moving positive charge is repelled with a greater electrostatic force the closer it gets to the fixed positive charge.

Section Five — Electrostatics

Explaining Electrostatic Phenomena

Detecting Charge: The Gold-leaf Electroscope

1) The basic components of a gold-leaf electroscope are a metal disc connected to a metal rod, at the bottom of which are attached two thin pieces of gold leaf.

2) When a charged object is pulled firmly across the disc, it transfers its charge through the metal rod to the gold leaves. Since like charges repel, the gold leaves rise.

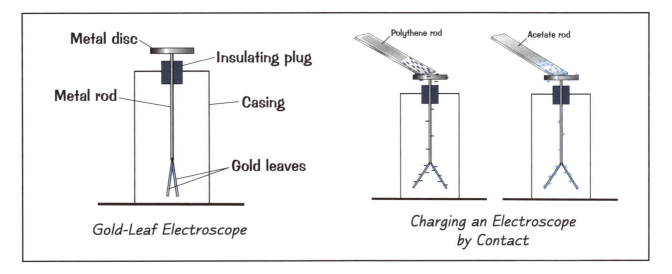

Gold-Leaf Electroscope

Charging an Electroscope by Contact

Now have a go at this question:

1) There is another way to charge an electroscope (other than by contact). Start off with an *uncharged* electroscope. See if you can predict what will happen when a charged object comes close to, but does not touch, the metal disc. [Hint: remember that electrons can move freely inside a metal]

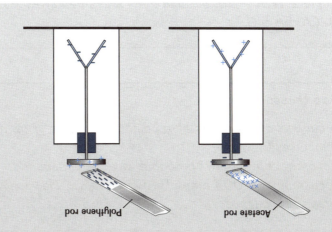

Answers

1) A positively charged rod attracts negatively charged particles onto the metal disc. Since the electroscope started off with equal amounts of positive and negative charge, this leaves the gold leaves positively charged; so the leaves repel each other. The same thing happens with a negatively charged object, but with the charges reversed.

Section Five — Electrostatics

SECTION SIX — CIRCUIT ELECTRICITY

Current

Electric Current — The Flow of Charge

When we connect a wire up to a battery, negatively charged particles (electrons) flow through the wire from the negative end of the battery to the positive end. We call this flow of charge an electric current, and, as you know, the charge will only flow if there is a complete circuit. The electric current at a point in the wire can be defined in the following way:

Current = the amount of charge passing the point / the time it takes for the charge to pass
(in amps)   (in coulombs)   (in seconds)

Or, equivalently, as the amount of charge passing the point in one second.

In symbols we can write this as:

$$I = \Delta Q / \Delta t$$

The "Δ" sign just means a small amount, so the equation reads: current equals the small amount of charge passing the point divided by the small amount of time it takes.

Here are a few examples:

1) 6 coulombs of charge flow through a lamp in one minute.
   What is the current through the lamp?
   $I = \Delta Q/\Delta t$, so $I = 6C/60s = 0.1A$

2) There is a current of 2 amps in a wire. How much charge will flow in 5 minutes?
   $I = \Delta Q/\Delta t$, multiplying both sides by Δt gives $I \times \Delta t = \Delta Q$, so $\Delta Q = 2A \times 300s = 600C$

3) There is a current of 0.5 amps through a resistor.
   How long does it take for 0.1 coulombs to pass through?
   From example 2, $I \times \Delta t = \Delta Q$, dividing both sides by I gives $\Delta t = \Delta Q/I$,
   so $\Delta t = 0.1C/0.5A = 0.2s$

Now have a go at these:

1) There is a current of 0.2 amps at a point in a circuit. How much charge will flow past that point in 2 minutes?
2) 0.01 coulombs of charge flow through a wire in 20 seconds. What is the current?
3) There is a current of 0.005 amps through a lamp. How long does it take for 1 coulomb of charge to pass through it?

Answers
1) 24C
2) 5×10^{-4}A (or 0.5mA)
3) 200s (or 3min 20s)

Current

What Happens When an Electric Current Meets a Junction?

We can easily build a circuit in which the electric current has a choice about which wire to travel down — two lamps connected in parallel is a good example.

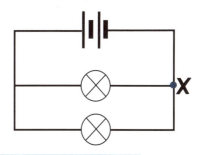

Consider what happens at point "X" — the current from the power supply can go down either of two possible routes. It can be difficult to work out precisely how much goes down each wire, but one thing is for certain:

> The total amount of charge leaving the junction every second must be the same as the total amount entering it every second.

Or, more concisely:

> The sum of the currents going into the junction = the sum of the currents going out.

This is a simplified statement of *Kirchoff's first rule*.

Consider the following examples with one unknown current.
In order to find the unknown current we simply use the rule:

1)
```
       0.5A
        ↑
  1A →──┼──→ 0.3A
        ↓
        I₁
```

1) Sum of currents in = sum of currents out
$1A = 0.5A + 0.3A + I_1$
$1A = 0.8A + I_1$
$1A - 0.8A = I_1$
$0.2A = I_1$

2)
```
       1.2A
        ↑
  1.5A→─┼──→ 0.7A
        ↑
        I₂
```

2) Sum of currents in = sum of currents out
$1.5A + I_2 = 1.2A + 0.7A$
$1.5A + I_2 = 1.9A$
$I_2 = 1.9A - 1.5A$
$I_2 = 0.4A$

Try some for yourself:

Find the unknown currents in the following three examples:

1)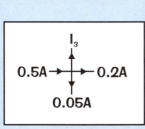

2)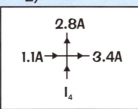

3)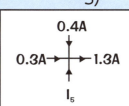

Answers:
1) $I_3 = 0.25A$
2) $I_4 = 5.1A$
3) $I_5 = 0.6A$

Section Six — Circuit Electricity

Voltage

Voltage — The Energy Per Unit Charge

In any circuit energy is transferred from the power supply to the components (lamps, motors etc.) where it is converted into other forms, light for example. This energy is carried by the charged particles. If we consider one coulomb of charge flowing around a circuit then:

(a) The amount of energy it is given by the power supply is the voltage across the power supply.
(b) The amount of energy it gives to each individual component in the circuit is the voltage across that component.

In other words, the voltage, or potential difference, across a component is the amount of energy (in joules) that it converts for every coulomb of charge that passes through it.

> Voltage across component = Energy converted / Charge that passes through it
> (in volts)   (in joules)   (in coulombs)
>
> Or, in symbols: $V = E/Q$

Here are a few examples:

1) A lamp gives out 10 joules of energy when 0.5 coulombs have passed through. What is the voltage across the lamp?
   $V = E/Q = 10J/0.5C = 20V$

2) What is the maximum amount of energy an electric heater could produce at 200 volts if the amount of charge that passes through it is 10 coulombs?
   $V = E/Q$, multiplying both sides by Q gives $V \times Q = E$, so $E = 200V \times 10C = 2000J$

3) How much charge has passed through a 12 volt motor if the energy it has converted is 3 joules?
   From example 2, $V \times Q = E$
   Dividing both sides by V gives $Q = E/V$, so $Q = 3J/12V = 0.25C$

Have a go at these problems:

1) What is the maximum amount of energy that a lamp could give out if the voltage across it is 6 volts and the amount of charge that has passed through it is 0.5 coulombs?
2) How much charge has passed through a circuit if 100 joules of energy have been converted across a voltage of 8 volts?
3) An electric motor converts 1 joule of energy when 0.04 coulombs of charge pass through it. What is the voltage across the motor?

Answers
1) 3J
2) 12.5C
3) 25V

Section Six — Circuit Electricity

Energy in Electrical Circuits

Conservation of Energy in Electrical Circuits

Energy is given to charged particles by the power supply and taken off them by the components in the circuit. Since energy is conserved, the amount of energy one coulomb of charge loses when going around the circuit must be equal to the energy it is given by the power supply.

Furthermore, this must be true regardless of the route the charge takes around the circuit. So, we can say that in most cases:

> For any closed loop in a circuit the sum of the voltages across the components must equal the voltage of the power supply.

This is a simple case of *Kirchoff's second rule*.

Have a look at this example:

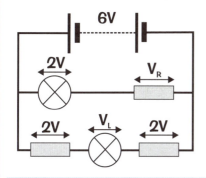

What is the voltage across the resistor, V_R, and the voltage across the lamp, V_L?

First look at just the top loop:

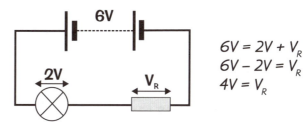

$6V = 2V + V_R$
$6V - 2V = V_R$
$4V = V_R$

Now look at just the outside loop:

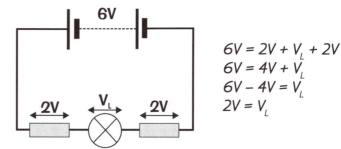

$6V = 2V + V_L + 2V$
$6V = 4V + V_L$
$6V - 4V = V_L$
$2V = V_L$

Now you do this one:

For the circuit on the right, calculate:
(a) the voltage across the motor, V_M
(b) the voltage across the loudspeaker, V_S.

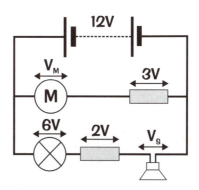

Answers:
(a) $V_M = 9V$
(b) $V_S = 4V$

Resistance

Resistance — The Ratio of Voltage / Current

Generally speaking, when there is a voltage across a component there will be a current through it. Usually, as the voltage is increased the current increases — this makes sense if you think of the voltage as a kind of force pushing the charged particles. We formalise the relationship between current and voltage by introducing the idea of "resistance":

Resistance of a component = Voltage across it / Current passing through it
    (in ohms)   (in volts)   (in amps)

Or, in symbols:

$$R = V / I$$

Multiplying both sides by I gives:

$$V = I \times R$$

Components with a <u>low resistance</u> allow a <u>large</u> current to flow through them, while components with a <u>high resistance</u> allow a only <u>small</u> current. Note that the resistance is not a constant — it may take different values as the current and voltage change.

Here are some examples:

1) If a voltage of 12 volts across a component causes a current of 0.001 amps to flow, what is the resistance?

   R = V/I, so R = 12V/0.001A = 12 000Ω, or 12kΩ

2) If a current of 0.2 amps passes through a resistance of 200 ohms, what is the voltage?

   V = I×R, so V = 0.2A×200Ω = 40V

3) What current will flow through a resistance of 800 ohms if the voltage across it is 6 volts?

   V = I×R, dividing both sides by R gives   I = V/R, so I = 6V/800Ω = 0.0075A = 7.5mA

Here are some for you to do:

1) If a current of 2.5 amps passes through a component with a resistance of 10 ohms, what is the voltage?
2) What current will flow through a resistance of 2500 ohms if the voltage across it is 6 volts?
3) What is the resistance of a component if 1.5 volts drives a current of 0.02 amps through it?

Answers
1) 25V
2) 0.0024A or 2.4mA
3) 75Ω

Resistance

Voltage-Current Graphs

Look at the following graphs showing the current through different components as the voltage across them is changed (negative values refer to charges flowing the other way):

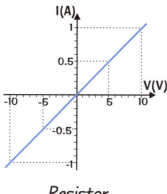

Resistor   Filament Lamp

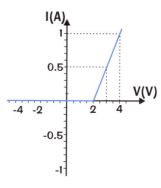

Diode

We can use the graphs to determine the resistance at different voltages as follows:

Example:
What is the resistance of the resistor at:
(a) -10V?
(b) -5V?
(c) 5V?
(d) 10V?

(a) R = V/I = -10V/-1A = 10Ω
(b) R = V/I = -5V/-0.5A = 10Ω
(c) R = V/I = 5V/0.5A = 10Ω
(d) R = V/I = 10V/1A = 10Ω

Now you have a go at the other two:

1) What is the resistance of the filament lamp at:
  (a) 10V?
  (b) 20V?
  (c) 30V?
  (d) 40V?

2) What is the resistance of the diode at:
  (a) 1V?
  (b) 2V?
  (c) 3V?
  (d) 4V?

Answers

1) (a) 25Ω
  (b) 28.6Ω
  (c) 33.3Ω
  (d) 40Ω

Note that the resistance of the lamp increases as the current and voltage increase.

2) (a) Infinite
  (b) Infinite
  (c) 6Ω
  (d) 4Ω

2V is the "breakdown potential" of this diode. Up to this voltage no current will flow, and no current will flow in the other direction for any voltage.

Section Six — Circuit Electricity

Power

Power — The Energy Converted Every Second

As we have said already, components in electrical circuits convert the energy carried by the charged particles into other forms; for example, a lamp converts it into light and heat energy. The amount of energy, in joules, that is converted every second is the power of that component:

Power (in watts) = Amount of Energy Converted (in joules) / Time Taken (in seconds)

Or, in symbols:
$$P = E/t$$

But (see p34) the energy converted is equal to the voltage across the component x the amount of charge that has flowed through it (E = V x Q).

So: $P = V \times Q / t$

and, the amount of charge that flows through a component is equal to the current through it x the time taken (Q = I x t)

So: $P = V \times I \times t / t$

Cancelling the "t"s gives: P = V x I.

Power (in watts) = Voltage Across Component (in volts) x Current Through Component (in amps)

Here are some examples:

1) An elevator motor converts 3×10^5 joules of electrical energy into gravitational potential energy (see pages 32-34) in a single one-minute journey. At what power is it working?

P = E/t, so P = 3×10^5 J/60s = 5000W, or 5kW.

2) If the voltage across a component is 6 volts and the current through it is 0.5 milliamps (5×10^{-4} amps), at what rate is it converting electrical energy to other forms?

P = I x V, so P = 5×10^{-4} A x 6V = 0.003W, or 3mW.

3) If a 12 watt lamp has a current through it of 1.5 amps, what is the voltage across it?

P = I x V, dividing both sides by I gives V = P/I, so V = 12W/1.5A = 8V.

Have a go at these questions:

1) What is the power output of a component if the current through it is 0.12 amps when the voltage across it is 6 volts?
2) What current passes through a 40 watt heater when the voltage across it is 10 volts?
3) How much electrical energy would be converted by the 40 watt heater in 10 seconds?

Answers
1) 0.72W
2) 4A
3) 400J

Section Six — Circuit Electricity

Power

Power — More Equations

We can combine the last equation for the power of an electrical component, $P = V \times I$, with the resistance equation (see p36) to create two more useful equations.

Firstly, replace V with I × R to give:

$$P = I \times R \times I = I^2 R$$

Power (in watts) = [Current (in amps)]^2 × Resistance (in ohms)

Secondly, replace I with V/R to give:

$$P = V \times V/R = V^2/R$$

Power (in watts) = [Voltage (in volts)]^2 / Resistance (in ohms)

Here are some examples — the key here is choosing the correct equation to use. If the question gives you the value of two variables and asks you to find another, you should choose the equation that relates these three variables. You may have to rearrange it before using it.

1) What is the power output of a component of resistance 100 ohms if the current through it is 0.2 amps?

   $P = I^2R$, so $P = (0.2A)^2 \times 100\Omega = 4W$. ($P = V^2/R$ would not have been any use here).

2) How much energy is converted in 10 s if the voltage is 100 V and the resistance is 5000 Ω?

   $P = V^2/R$, so $P = 100^2/5000 = 2W$. But power is the energy converted per second, so the energy converted in 10s is $2W \times 10s = 20J$. (Again, $P = I^2R$ would not have helped much)

3) Resistors get hotter when a current flows through them. If you double the current through a resistor, what happens to the amount of heat energy produced every second?

   It increases by a factor of 4 — this is because the current is squared in the expression for the power (try substituting some values of I and R to confirm this).

4) If a lamp has a power rating of 6 W and the voltage across it is 12 V what is the resistance?

   Now, we do not have an expression for the resistance so far, but we know which equation to use because we're given the power and the voltage. So we need to use the equation that relates P, V and R, namely $P = V^2/R$. Multiplying both sides by R gives $P \times R = V^2$, and dividing by P gives:

   $R = V^2/P$, so $R = (12V)^2/6W = 24\Omega$

Now have a go at these:

1) What is the power output of a component of resistance 2000 ohms if the current through it is 1.2 amps?
2) How much energy is converted in 1 minute if the resistance is 100 Ω and the current is 2 A?
3) If a lamp has a power rating of 6 watts and the current through it is 0.5 amps, what is the resistance?

Answers: 1) 2880W   2) 24 000J, or 24kJ   3) 24Ω

SECTION SEVEN — RADIOACTIVITY

Nuclear Radiation

The Constituents of the Atom

The diagram below is a representation of a lithium atom:

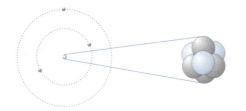

Where:

- = neutron, relative mass 1, charge 0
- = proton, relative mass 1, charge +1
- = electron, relative mass 1/1840, charge −1

We write the symbol for this atom: $^{7}_{3}Li$

where 3 is the number of protons in the nucleus (the proton number) and 7 is the total number of protons and neutrons in the nucleus (the atomic number).

Three Kinds of Nuclear Radiation

There are three kinds of nuclear radiation you need to know about.

Alpha Radiation (Symbol α)

An alpha particle is emitted from the nucleus:

It is made up of two protons and two neutrons.

As a result, the proton number of the original atom goes down by 2 and the atomic number goes down by 4. E.g. The alpha decay of Radium-226:

$$^{226}_{88}Ra \rightarrow \,^{222}_{86}Rn + \,^{4}_{2}\alpha$$

Beta Radiation (Symbol β)

A neutron in the nucleus turns into a proton and an electron — the electron is emitted from the nucleus and is called a beta particle. As a result the proton number of the nucleus goes up by 1, but the atomic number does not change. E.g. The beta decay of Radium-228:

$$^{228}_{88}Ra \rightarrow \,^{228}_{89}Ac + \,^{0}_{-1}\beta$$

Nuclear Radiation

Gamma Radiation (Symbol γ)

High energy electromagnetic radiation, known as gamma radiation, is emitted from the nucleus. The number of protons and neutrons in the nucleus stays the same.
E.g. The gamma decay of Iodine-131:

$$^{131}_{53}I \rightarrow {}^{131}_{53}I + {}^{0}_{0}\gamma$$

> N.B. When a nucleus emits any of these forms of radiation we call it decay.
> By decaying, a nucleus reduces its energy, making it more stable.

Have a go at these questions:

1) State how many protons and how many neutrons there are in each of the following nuclei:

   a) $^{241}_{95}Am$ b) $^{239}_{94}Pu$ c) $^{90}_{38}Sr$ d) $^{60}_{27}Co$ e) $^{226}_{88}Ra$

2) Copy and complete the following decay equations, making sure that you put the correct proton and atomic numbers for the nuclei:

   $^{242}_{94}Pu \rightarrow {}^{\_\_}_{\_\_}U + {}^{4}_{2}\alpha$

   $^{\_\_}_{\_\_}K \rightarrow {}^{40}_{20}Ca + {}^{0}_{-1}\beta$

   $^{222}_{86}Rn \rightarrow {}^{218}_{84}Po + {}^{\_\_}_{\_\_}$

   $^{60}_{27}Co \rightarrow {}^{\_\_}_{\_\_} + {}^{0}_{0}\gamma$

   $^{14}_{6}C \rightarrow {}^{\_\_}_{\_\_}N + {}^{0}_{-1}\beta$

Answers

1) (a) 95, 146
   (b) 94, 145
   (c) 38, 52
   (d) 27, 33
   (e) 88, 138

2) $^{242}_{94}Pu \rightarrow {}^{238}_{92}U + {}^{4}_{2}\alpha$

   $^{40}_{19}K \rightarrow {}^{40}_{20}Ca + {}^{0}_{-1}\beta$

   $^{222}_{86}Rn \rightarrow {}^{218}_{84}Po + {}^{4}_{2}\alpha$

   $^{60}_{27}Co \rightarrow {}^{60}_{27}Co + {}^{0}_{0}\gamma$

   $^{14}_{6}C \rightarrow {}^{14}_{7}N + {}^{0}_{-1}\beta$

Section Seven — Radioactivity

The Random Nature of Radioactive Decay

Radioactive Decay is a Random Process

Consider a sample of 10 000 radioactive nuclei that have not yet decayed.
It is not possible to predict which nucleus will decay next, or when (decay is random).
We can, however, consider the probability of a nucleus decaying.

Let's say that the probability of any given nucleus decaying
(and emitting α, β or γ radiation) in the next second is 1 in 10.
It is then a fair bet that after one second roughly 1000 of those
10 000 nuclei will have decayed, leaving about 9000 undecayed nuclei.
How about the next second?
There are now only 9000 undecayed nuclei left, so roughly 900 of them will decay in the next
second, leaving 8100 undecayed after two seconds (roughly speaking).

If we continue along these lines we can plot a graph of the number of undecayed nuclei left as time goes on:

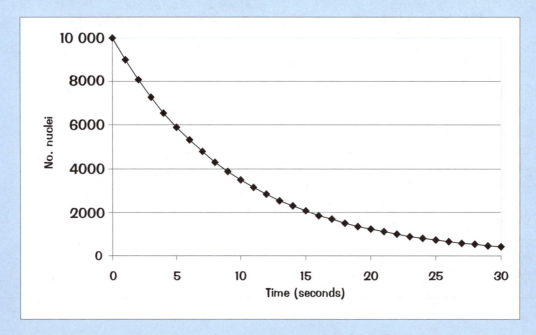

Half-Life

The graph above is a special kind of curve called an exponential decay.
It has a feature which we call the half-life.

The half-life is the time for the number of undecayed nuclei to fall to half the original number.

If we wait for 2 half-lives the number will fall to a quarter of its original value,
if we wait for 3 half-lives the number will be an eighth, 4 half-lives a sixteenth etc.

The number of undecayed nuclei falls by a factor of two every half-life.

The half-life for our example is about 7 seconds.

Section Seven — Radioactivity

The Random Nature of Radioactive Decay

Consider the following examples:

1) By what factor will the number of undecayed nuclei have fallen after 8 whole half-lives?

> The number of nuclei halves every half-life, so after 8 it should be:
> ½ × ½ × ½ × ½ × ½ × ½ × ½ × ½ = (½)^8 = 1/256.

2) How many half-lives does it take for the number to have fallen by a factor of 1024?

> ½ × ½ × ½ × ½ × ½ × ½ × ½ × ½ × ½ × ½ = 1/1024
> i.e. 10 half-lives.

Have a go at these questions:

1) After how many half-lives will the number of undecayed nuclei have fallen by a factor of 64?
2) A sample of 8 million nuclei have a half-life of 3 minutes. How long does it take for the number of undecayed nuclei remaining to have fallen to 1 million?
3) Consider 2 samples of radioactive nuclei:
   A: Starts with 64 million nuclei and has a half-life of 2 days.
   B: Starts with 16 million nuclei and has a half-life of 3 days.
   How long will it be until both samples have the same number of undecayed nuclei?

Answers
1) 6
2) 9 minutes
3) 12 days (they will both have 1 million nuclei left undecayed.)

Section Seven — Radioactivity

Index

A
acceleration 12, 13, 16, 18, 19, 20
acceleration due to gravity 13
Alpha Radiation 40
amplitude 1, 2, 5
atom 40
atomic number 40

B
balanced forces 20
battery 32
Beta Radiation 40
braking force 18

C
catapulted 23
change in velocity 11
charge 29, 30, 31, 32, 33, 34, 35, 38
circuits 32, 33, 34, 35
conservation of energy 23, 35
constant velocity 27
coulomb 29, 34, 35
current 32, 33, 36, 37, 38, 39

D
decelerating 16
diode 37
direction of motion 10
displacement 2, 8, 9, 10, 14
displacement-distance graphs 1, 2, 14, 15
distance 6, 7, 14, 16, 24
distance-time graphs 14, 15

E
efficiency 28
electromagnetic waves 4
electrons 32, 40
Electroscope 31
electrostatic force 29, 30
electrostatic phenomena 30, 31
energy 1, 21, 22, 23, 24, 25, 26, 28, 34, 35, 38
equilibrium position 1
exponential decay 42

F
falling 13, 23
filament lamp 37
forces 18, 19, 20, 24, 27
frequency 3, 4, 5

G
Gamma Radiation 41
gold-leaf electroscope 31
gradient 14, 16
gravitational field strength 22, 25
gravitational potential energy 21, 22, 23, 25, 38
gravity 13, 29

H
Half-Life 42
height 22
hertz (Hz) 3

J
junction 33

K
kinetic energy 21, 22, 23, 25, 26, 28
Kirchoff's first rule 33
Kirchoff's second rule 35

L
longitudinal waves 1

M
mass 18, 19, 21, 22, 29
motor 28

N
net force 20
neutron 40
Newton's Second Law 18

O
Ohm's law 36, 39
oscillation 2, 3
output 26

P
parallel 33
particles 29, 30
potential difference 34
potential energy 28
power 26, 27, 38, 39
proton 40
proton number 40
Pythagoras 8

R
radiation 42
radio waves 4
radioactive decay 42
resistance 36, 37, 39
resistor 37, 39
resultant 8
resultant force 20

S
scale drawing 8, 11
slinky 1, 3
speed 4, 6, 7, 14, 15, 16, 21
speed-time graphs 16, 17

T
time 6, 7, 14, 16, 26, 27, 32, 38
time period 2, 3, 5
tip-to-tail 8, 20
transverse waves 1

U
unbalanced forces 20

V
vector 8, 9, 11, 12, 20
velocity 10, 11, 12, 14, 15, 18
velocity-time graphs 16, 17
vibrating 3
vibrations 1
voltage 34, 35, 36, 37, 39
voltage-current graphs 37

W
wave equation 4, 5
wave speed 5
wavelength 1, 4, 5
waves 1, 2, 3, 4, 5
weight 22
work done 24, 25, 26, 27, 28